Edward Muñoz Garro

Global Leadership in the Digital Age

Edward Muñoz Garro

Global Leadership in the Digital Age

AI, Transformation and Strategies for Innovation

ScienciaScripts

Imprint

Cover image: www.ingimage.com

This book is a translation from the original published under ISBN 978-620-0-02619-4.

Publisher:
Sciencia Scripts
is a trademark of
Dodo Books Indian Ocean Ltd. and OmniScriptum S.R.L publishing group

120 High Road, East Finchley, London, N2 9ED, United Kingdom
Str. Armeneasca 28/1, office 1, Chisinau MD-2012, Republic of Moldova, Europe
Managing Directors: Ieva Konstantinova, Victoria Ursu
info@omniscriptum.com

Printed at: see last page
ISBN: 978-620-8-60113-3

Global Leadership in the Digital Age

AI, Transformation and Strategies for Innovation

By Edward Muñoz Garro

Table of Contents

Foreword

We live in an era in which digital transformation has become a fundamental axis for defining the strategies, culture and competitiveness of organizations. Throughout my professional career as a Project Manager and specialist in technology and finance, I have had the opportunity to collaborate with various teams and sectors that, faced with technological disruption, have had to rethink the way they operate, create value and relate to their collaborators. This book arises precisely from this combination of experiences, formal studies and deep reflections that I have accumulated in projects of modernization and adoption of artificial intelligence, automation and digitization of processes.

When I started my career, the ideal of digital transformation was still perceived as a distant, almost futuristic aspiration. However, the acceleration of the Fourth Industrial Revolution has made adopting cloud solutions, implementing agile methodologies and leveraging big data an inescapable responsibility for today's leaders. I have experienced firsthand how, in high-pressure environments and global competition, technology is not just a tool, but a catalyst that redefines the work of the organization, opens business opportunities and, at the same time, challenges the human factor.

Throughout my work in specialized consulting -including financial automation and cloud modernization projects-, I have observed that the success of digital transformation does not depend exclusively on algorithms or technological infrastructure, but on the vision and commitment of those who lead. In my experience, those who manage to align project management methodologies with business culture and financial goals are able to turn the uncertainties of change into a driver of competitiveness. This involves integrating elements such as emotional intelligence, financial negotiation and rigorous risk management.

In these pages, it reviews how artificial intelligence, automation and the adoption of innovations transform global leadership in digitized contexts . Beyond the technical perspective, the text delves into theoretical models-such as Everett Rogers' model for innovation diffusion, Kurt Lewin's approach to managing change, and Bernard Bass' transformational vision-that provide a solid framework for understanding and guiding the impact of technology on organizational dynamics. Special emphasis is placed on the intersection between culture, resistance to change and strategic decisions, for it is there that I have found that the success or failure of innovation is woven.

My intention with this book is, first, to provide an academic and up-to-date analysis of the factors underlying digital transformation; and second, to offer practical reflections based on my own experience in project management and technical-financial consulting. Therefore, each chapter is built on the premise that technology is inseparable from the people who implement it and the culture that embraces it. In today's dynamic and demanding market, we need leaders with a global outlook, able to navigate between innovation and empathy, efficiency and ethics, speed and continuous training.

I invite you to dive into these pages with an analytical and curious attitude, because digital transformation, well managed, can become one of the greatest levers of growth and differentiation. You will find discussions on cultural adaptation, building shared visions and managing geographically dispersed teams. Also covered are the challenges and opportunities that AI and automation bring, from process optimization to reinventing job roles. My hope is that this book will not only explain the changes underway, but inspire the leaders of today - and tomorrow - to lead their organizations with skill and humanity in the digital age.

Because project management and innovation are not an end in themselves, but a path towards more agile, inclusive and sustainable organizations. To that end, I offer here a critical and constructive vision of digital transformation, supported by academic research and field experiences. I thank in advance those who join me on this journey, convinced that technology and leadership go hand in hand when we seek authentic progress.

Edward Muñoz Garro, MBA

Project Manager - Technology and Finance Expert

Chapter 1. Introduction and Context

Digital transformation has become one of the most disruptive and relevant phenomena for organizations today. The accelerated implementation of technologies such as artificial intelligence (AI), automation and big data analysis is not only redefining business processes, but is also substantially changing the way leaders approach strategy, organization and corporate culture. Specifically, in an increasingly globalized and competitive world, digital leadership is no longer optional: it has become a key axis for sustainability and innovation.

This chapter provides an overview of the relevance of global leadership in highly digitized environments. It describes why digital transformation is an urgent priority for organizations of all kinds, outlines the objectives guiding this book and its methodological framework, and presents the overall structure of the book. At the end of this chapter, the reader will have a clear context on the dynamics between resistance to change, leadership and digital transformation, which will be analyzed in the following chapters.

1.1 Justification of the subject

1.1.1 Importance of digital transformation in global leadership

In today's changing business world, technology has emerged as a determining factor in the competitiveness of organizations and the performance of their leaders. According to Deloitte Insights (2020), managers who drive the adoption of emerging technologies tend to generate higher levels of innovation and resilience in their companies, especially in uncertain environments. On the other hand, reports such as the World Economic Forum's "Global Technology Governance Report 2021" (2021a) highlight how digitalization, accelerated after the COVID-19 pandemic, has altered not only the internal workings of companies, but also the nature of work and the dynamics of collaboration.

From this perspective, global leadership in the digital era is not just about understanding technology, but also involves the ability to move resources and teams across multiple geographies and cultures, promoting the adoption of technological tools and fostering the integration of diverse perspectives. The rise of automation and AI has also generated an ongoing debate about the role of emotional intelligence, creativity and ethics in decision making. While the traditional leader focused primarily on task oversight, the digital leader is forced to integrate technological competencies and advanced interpersonal skills to ensure the consistency of his or her strategic vision.

Likewise, Cortellazzo et al. (2019) compare the disruption caused by digitization to the revolution brought about by the movable printing press at the time. Just as the printing press democratized information and transformed access to knowledge, today's digitization redefines the way we communicate, produce and collaborate. In this context, the figure of the leader becomes relevant by directly influencing the organizational culture and the willingness of teams to innovate.

1.1.2 Relevance in the Fourth Industrial Revolution

The so-called Fourth Industrial Revolution, or Industry 4.0, is characterized by the convergence of physical, digital and biological technologies, leading to transformations of great magnitude in the value chains of organizations (World Economic Forum, 2021b). Among the agents of change, artificial intelligence, advanced robotics, the internet of things (IoT), machine learning and big data analytics stand out. Beyond the technical aspects, this phenomenon raises ethical, social and human challenges.

As Mendenhall et al. (2012) point out, global leaders are challenged to lead multicultural and geographically dispersed teams. In addition to these requirements, there is now the need to understand and manage the adoption of technological tools that accelerate processes, but at the same time generate uncertainty and resistance in the staff. In addition, the deficit of digital competencies becomes an obstacle to the effective implementation of the transformation, requiring training and support efforts by leaders.

Consequently, the relevance of leadership in the Fourth Industrial Revolution transcends the simple adoption of innovations: it implies a change in organizational mindset that weaves agility, inclusion, sustainability and ethical transparency into corporate strategy. This scenario substantiates the need to deepen the relationship between digital transformation and global leadership and justifies the detailed study of the factors that promote or hinder the effective adoption of technology.

1.2 Objectives and scope of the work

1.2.1 General objective and specific objectives

The purpose of this book is to take an in-depth look at how artificial intelligence, automation and other emerging technologies influence the redefinition of global leadership in digitized business environments. From this general objective, several specific objectives have been delineated to guide the discussion:

1. **To examine** how digital technologies expand the skills and capabilities of leaders, modifying their roles and management methods.
2. **Assess** the impact of AI and automation on strategic decision making, considering both benefits (efficiency, accuracy) and risks (dehumanization, algorithmic dependency).
3. **Identify** the challenges and resistances that arise when implementing significant technological changes, exploring strategies to overcome them based on theories of organizational change and innovation adoption.
4. **Propose** guidelines and best practices that integrate digital competencies, emotional intelligence and ethical leadership in the age of automation and big data.

This approach seeks to provide a critical and practical vision for managers, consultants, students and researchers who wish to understand how to capitalize on the opportunities of digitalization while mitigating its potential adverse effects on human capital and organizational culture.

1.2.2 Research questions

To align the specific objectives with the research, the following key questions are formulated:

- How do AI and automation influence the distribution of responsibilities and the definition of leadership roles?
- How is emotional intelligence integrated into business decisions as data analytics and algorithms become more prominent?
- Which leadership models have proven most effective in contexts of accelerated digital transformation and why?
- How can resistance to change be overcome in organizations where digitalization threatens the stability of some job roles?
- What are the ethical and social responsibility implications of automating processes and delegating decisions to AI systems?

These questions will serve as the guiding theme throughout the book, linking the theoretical framework with the case studies and the final reflections.

1.3 Structure of the book

To achieve the stated purposes, the book is organized into chapters that combine theoretical foundation, critical discussion and practical application:

1. **Chapter 1. Introduction and Context** Presents the rationale for the topic, the importance of digital transformation in global leadership, the objectives and research questions. It also includes the methodological approach adopted.
2. **Chapter 2. Theoretical Foundations and Key Models** Delves into the main conceptual frameworks that support the study: Everett Rogers' Innovation Adoption Model , Kurt Lewin's Organizational Change Theory and Bernard Bass' Transformational Leadership Model. It examines how these theories provide an understanding of technology diffusion, change management and the role of the leader as a transformational agent.
3. **Background on Digital Transformation and Global Leadership** Explores the historical evolution of digitalization, recent trends identified by international consulting firms and institutions (Deloitte Insights, World Economic Forum), and how leaders have addressed emerging challenges.

It justifies why it is critical to analyze these changes from a leadership perspective.

4. **Chapter 4. Artificial Intelligence and Automation: Impacts on Decision Making** Analyzes in greater detail how AI and automation alter the decision process in organizations, highlighting both its benefits and potential risks. Examples of sectors where machines have already replaced or complemented human intervention are addressed.
5. **Chapter 5. Digital Leadership Skills and Adaptability** Identifies the digital and relational competencies that enable the leader to take advantage of technological tools without losing sight of the relevance of the human dimension. It highlights the importance of emotional intelligence, virtual communication and adaptability.
6. **Chapter 6. Resistance to Change and Organizational Transformation** Examines resistance to change, one of the most frequent barriers to digitalization. The phases of unfreezing, change and refreezing proposed by Kurt Lewin are taken as a reference, as well as concrete strategies to manage the uncertainty and fear that technological adoption can generate.
7. **Chapter 7. Integrating People and Technology: Hybrid Leadership** Delves into the need to achieve a balanced convergence between technological and human competencies, emphasizing the role of empathy, ethics, creativity and shared vision.
8. **Chapter 8. Strategies and Tools for Digital Transformation** Provides models, methodologies and practical guides to drive and evaluate digital transformation initiatives, addressing cybersecurity, continuous training and organizational culture.
9. **Chapter 9. Case Studies and Lessons Learned** Presents examples of companies, such as General Electric and Ford, that have implemented massive digital transformations. Their successes and failures are analyzed in light of the conceptual frameworks presented.
10. **Chapter 10. Future Challenges and Emerging Trends** Reflects on the areas of opportunity and threats looming in the medium and long term,

including collaborative robotics, algorithmic ethics, the metaverse, and the new formative demands of digital leadership.

11. **General Conclusions and Recommendations** Recapitulates the most relevant findings and proposes concrete lines of action for managers, researchers and decision makers seeking to align digital transformation with sustainable organizational development.

1.4 Methodological approach

1.4.1 Scope and perspective of the investigation

This paper has been developed under a qualitative approach and a theoretical scope, oriented to analyze how artificial intelligence and automation are redefining global leadership in highly digitized organizations. Emphasis is placed on the literature review and the analysis of real cases, in order to identify common patterns in the adoption of technological innovations and in the management of organizational change.

1.4.2 Units of analysis and sources consulted

Units of analysis included:

- Academic papers (scientific articles, book chapters) and institutional publications (Deloitte reports, World Economic Forum).
- Case studies of companies that have undergone digital transformation processes, with those of General Electric and Ford standing out for their scope and complexity.
- Fundamental theoretical references: Everett Rogers' Innovation Adoption Model, Kurt Lewin's Organizational Change Theory and Bernard Bass' Transformational Leadership Model.

The main databases for literature location were JSTOR, PubMed and Google Scholar, applying search equations that combined key terms such as "global leadership", "artificial intelligence", "digital transformation" and "resistance to change". In addition, citation and relevance filters were used to prioritize recent studies.

1.4.3 Data Collection and Processing Techniques

1. **Systematic literature review**: documents that directly or tangentially addressed the relationship between technology and leadership were identified and analyzed, along with reports from international organizations and prestigious consulting firms.
2. **Qualitative content analysis**: an inductive approach was adopted to classify the findings and relate them to the selected theoretical models.
3. **Contrast with case studies**: reports and experiences of companies in the process of digitalization were reviewed, focusing on the leadership and change management strategies applied.

This methodological procedure made it possible to draw conclusions based on various sources, ensuring both theoretical depth and practical relevance of the results.

Chapter 2. Theoretical Foundations and Key Models

Digital transformation cannot be fully understood without analyzing how innovations are adopted in organizations, how change processes are managed, and what role leadership plays in motivating and guiding teams. To this end, this chapter delves into three theoretical models that provide a robust analytical framework: Everett Rogers' Innovation Adoption Model, Kurt Lewin's Organizational Change Theory and Bernard Bass' Transformational Leadership Model. These approaches allow us to understand, from different angles, how technology influences organizational dynamics and how leaders can successfully orchestrate significant change.

2.1 Everett Rogers' Innovation Adoption Model

2.1.1 Origins and fundamentals

The Innovation Adoption Model proposed by Everett Rogers (2003) has been widely recognized as one of the cornerstones for understanding the diffusion of new ideas, products or technologies in different social and organizational environments. Rogers proposes that the adoption of innovations occurs through a process in which different segments of the population -or of the organization- (innovators, early adopters, early majority, late majority and laggards) present differentiated attitudes and behaviors when adopting a novelty.

This approach highlights factors such as:

- **Relative advantage**: how much potential users perceive that the innovation offers superior benefits to previous practices or technologies.
- **Compatibility**: the extent to which the innovation fits with the values, experiences and needs of the organization.
- **Complexity**: how easy or difficult it is to understand and use the new technology or procedure.
- **Observability**: how visible the usefulness of the innovation is to others.
- **Trialability**: the possibility of testing the innovation on a small scale before widespread deployment.

2.1.2 Critical Contributions and Reviews

Rogers et al. (2005) and subsequent studies argue that organizations can be considered complex adaptive systems, in which the diffusion of innovations responds not only to formal communication, but also to informal interactions, shared perceptions and internal cultural dynamics. In this sense, leadership plays a key role in encouraging or inhibiting adoption by defining strategic priorities, allocating resources and shaping attitudes towards innovation.

- **Vanderslice (2000)** critically reviews how Rogers' extensive synthesis can be applied to multiple academic fields, from pedagogy to project management.
- **Hoffmann (2007)**, for his part, offers a retrospective look at the five editions of the original work, emphasizing how Rogers evolved from an initial sociological emphasis toward consideration of systems complexity and patterns of resilience.

2.1.3 Relevance for digital transformation

In the context of digital transformation, Rogers' model helps to understand why certain organizations quickly adopt technologies such as artificial intelligence and automation, while others are skeptical or delay the process. It also explains how resistance to change can emerge if the innovation is perceived as complex, incompatible with the existing culture or of uncertain benefit.

Furthermore, van Oorschot et al. (2018) note that the adoption of digital innovations is affected by factors such as technological infrastructure, willingness to invest in training, and the strategic vision of top management. Therefore, the leader's job is to facilitate the diffusion process, by allocating resources, promoting controlled experimentation, and transparently communicating the expected benefits of the innovation.

2.2 Kurt Lewin's Theory of Organizational Change

2.2.1 Conceptual structure: defrosting, switching and refreezing

Kurt Lewin's Theory of Organizational Change is one of the pioneering approaches to explain how organizations move from a current state to a desired future state. Lewin (1951) describes change as a three-phase process:

1. **Unfreezing**: The status quo is questioned and an awareness of the need for change is created, breaking entrenched routines.
2. **Change (or transition)**: New practices, values or technologies are introduced, requiring the involvement of people and the acquisition of different competencies.
3. **Refreezing**: The innovation or new state is consolidated and stabilized, integrating itself into the culture and structures of the organization.

While Lewin's model has received criticism for being considered linear and not reflecting the complexity of today's environments, Endrejat & Burnes (2024) argue that its fundamentals remain valid in the Fourth Industrial Revolution, as they provide a basic framework for understanding the dynamics of resistance and acceptance, as well as the importance of continuous reinforcement and feedback.

2.2.2 Application in digital transformation

In digital transformation processes, unfreezing involves making the organization aware of the urgency of adopting AI or automation. Often, resistance lies in fear of job displacement or insecurity about the skills required. The leader must articulate why the change is needed and what the strategic purpose of technology adoption is.

During the change stage, training, fluid communication and the active participation of employees are essential. Here, the leader who applies transformational leadership, listening to his teams and resolving concerns, will tend to achieve greater engagement. Finally, the refreezing phase refers to the stabilization of digital practices, the creation of protocols and the constant evaluation of results, so that the new behavior becomes part of the corporate culture.

2.2.3 Synergy with other models

When Lewin's theory is combined with Rogers' model, it is possible to understand both the organizational change process and the diffusion of innovation at the individual and collective levels. Thus, the leader can design strategies adjusted to the stage of adoption in which employees find themselves, while using Lewin's phases to manage transitions and reduce resistance.

2.3 Bernard Bass' Transformational Leadership Model

2.3.1 Concept and evolution

The Transformational Leadership Model, first proposed by James MacGregor Burns (1978) and further developed by Bernard Bass (1985), describes a leadership style in which the leader inspires, motivates and encourages followers to identify with and commit to a transcendent vision. Bass and Avolio (1994) detail the four fundamental components of this type of leadership:

1. **Charisma or idealized influence**: The leader projects integrity and trust, becoming a moral and professional reference.
2. **Inspirational motivation**: The leader articulates an attractive and compelling vision, generating enthusiasm and a sense of collective purpose.
3. **Intellectual stimulation**: Encourages creativity and critical thinking, challenging assumptions and promoting innovation.
4. **Individualized consideration**: Paying attention to the personal needs and aspirations of each employee, offering mentoring and guidance.

2.3.2 Relevance in digital environments

In the era of digital transformation, the transformational leader takes on a pivotal role, for:

- **Drives change**: By projecting a clear vision of how technology can improve efficiency and competitiveness, it mobilizes the organization to adopt innovation.
- **Manages uncertainty**: Through inspirational motivation and individualized consideration, mitigates fear of job replacement or disruption of traditional processes.
- **Promotes creativity and experimentation**: If employees feel supported, they are more willing to take risks, propose improvements and adapt to technological changes.

According to Deng et al. (2023), transformational leadership is positively related to attitude toward AI adoption, as inspirational communication and empathy facilitate the acceptance of digital solutions. Furthermore, Bass and Riggio (2006) argue that highly complex environments require leaders who integrate the human aspect with technological innovation.

2.3.3 Integration with innovation adoption and change management

The combination of the Transformational Leadership Model with Rogers' and Lewin's models provides a holistic perspective of digital transformation processes. Transformational leadership acts as a catalyst for innovation (Rogers) to spread faster and for people to move through the stages of change with less resistance (Lewin). Thus, the leader not only introduces technology, but also promotes a favorable environment for experimentation and co-creation.

2.4 Convergence of Models and their Relevance in the Digital Era

The Rogers, Lewin and Bass models do not operate in isolation. On the contrary, each covers a different level of the digital transformation phenomenon:

- **Individual and collective diffusion of innovation (Rogers)**: It explains how and why individuals and groups decide to adopt or reject a technology.

- **Organizational change management (Lewin)**: Emphasizes the macro processes of cultural and structural transition, illustrating how change is implemented and consolidated in the organization.
- **Role of the leader as a facilitator of change and innovation (Bass)**: Highlights the qualities and strategies that leaders must deploy to align people with the vision and generate deep commitment.

In digital environments, the synergy of these three approaches facilitates understanding and managing complex scenarios: from the first phase of awareness of the need to transform, to the consolidation of new technologies in the organizational culture. Leadership emerges as the common thread that favors the smooth adoption of innovation (Rogers), the effective management of the change process (Lewin) and constant inspiration and motivation (Bass).

2.5 Chapter Reflections

A review of the theoretical foundations of Rogers, Lewin and Bass reveals that, behind any technological or disruptive leap, there are human and cultural dynamics that dictate the real pace of change. It is not just a matter of introducing an innovation or redefining processes; the real difference lies in how leadership mobilizes - or not - collective energies towards a new reality.

This chapter reminds us that resistance to change, a phenomenon that often frustrates managers, is not an insurmountable obstacle but a sign that greater empathy, better communication and a shared vision are needed. And it is precisely here that transformational leadership, grounded in inspiration and individualized consideration, can bring to fruition the motivation of people who did not see themselves as innovators.

As we move into the Fourth Industrial Revolution, one certainty emerges: the success of digital transformation depends as much on the technical robustness of a tool as it does on the leader's ability to foster curiosity and engagement among his or her team. Ultimately, when we talk about spreading AI or automation, we are talking about cultivating the trust of those who will interact with these technologies on a day-to-day basis, whether in the boardroom or on the production line.

Therefore, this theoretical part, far from being a simple review of conceptual frameworks, becomes a map for each leader or professional to recognize where his or her organization is and what type of intervention is needed. Looking ahead to the next chapters, the invitation is to reflect:

- How are the phases of change being managed in my environment?
- At what stage of innovation adoption are my employees?
- In what ways am I fostering - or limiting - the creativity and confidence necessary for experimentation?

The answers to these questions outline the roadmap for turning digital transformation into a catalyst for opportunity, and not merely a technological imposition. In the following chapters, we will see how this dynamic has been experienced in real cases and what concrete strategies can be employed to align business vision with human development and the infinite potential of innovation.

Chapter 3. Background on Digital Transformation and Global Leadership

Today, digital transformation has become a continuous process that integrates the adoption of disruptive technologies, the redesign of business models and cultural transformation within organizations. Understanding its historical roots, the drivers of the Fourth Industrial Revolution and the resulting landscape of opportunities and challenges is vital to explain how global leadership is being forced to reinvent itself and take on increasingly complex competencies.

3.1 From Early Automation to AI: A Brief Historical Evolution

3.1.1 Early Automation and Process Electrification

Initial business digitalization efforts emerged in the mid-20th century, when early automation focused on replacing repetitive manual tasks with electrical and mechanical mechanisms, mainly in the industrial field. The advent of robots on production lines reduced errors and costs, ushering in a new era of optimization in manufacturing sectors.

- **Classic Example**: The adoption of robotic arms in the automotive industry allowed the production of large volumes with uniformity and greater quality control.
- **Limitations**: These systems offered little flexibility and tended to require very high investments, creating a gap between the production floor - highly automated - and the support or administrative areas - less digitized.

3.1.2 Computerization and Emergence of Networks

The advent of mainframe computers and the expansion of local area networks (LAN) marked a milestone in the computerization of several corporate areas: accounting, payroll, inventories, among others. This laid the groundwork for massive digitalization, as databases and internal processes began to be consolidated in enterprise resource planning (ERP) systems.

- **Professionalization of Roles**: System administrators and data analysts appear, who maintain and manage corporate information.

- **Strategic Focus**: Computers are no longer simple computing machines, but nodes of **communication and integration**, paving the way for interdisciplinary collaboration and real-time data flow.

3.1.3 Digital Convergence: Internet and Web Services

The 1990s and the early years of the new century were marked by the explosion of the Internet and the proliferation of web services. Global access to information and continuous connectivity enabled new forms of collaboration, the emergence of digital business models (e-commerce, SaaS, etc.) and a greater understanding of technology as a key factor in competitiveness.

- **Value Transformation**: Customer experience and continuous innovation became key differentiators.
- **Cultural Changes**: Large and small organizations began to bet on flatter and more flexible structures, oriented to the integration of technology in all areas.

3.1.4 The Rise of Artificial Intelligence

With the rise of cloud computing and the availability of machine learning algorithms, AI expanded beyond academic research, entering corporate and government domains. Companies discovered the predictive and analytical power of AI, leading to significant advances in personalized recommendations, fraud detection and logistics process optimization.

- **Deep Change**: AI not only streamlines tasks, but analyzes complex patterns, anticipates scenarios and reduces uncertainty.
- **Leadership Requirements**: Integrating AI into business strategy requires leaders capable of interpreting the technology and guiding the development of competencies in the workforce, promoting adaptation and innovation.

3.2 Drivers of the Fourth Industrial Revolution

The so-called Fourth Industrial Revolution is defined by the interconnectedness of technologies such as the internet of things (IoT), collaborative robotics, cloud computing and artificial intelligence. According to the World Economic Forum (2021), these are some of the key factors:

- Massive Access to Technological Infrastructure
  - Cloud computing and usage-based subscriptions have made entry to cutting-edge technologies cheaper, allowing both SMEs and large corporations to adopt them.
- Extreme Datafication
  - The collection and analysis of data in gigantic volumes (big data) has become the backbone of strategic decision making.
- Evolution of Robotics and Automation
  - Increasingly intelligent robots, capable of working in teams with humans and adapting quickly to changes in demand or the environment.
- Need for Resilience and Agility
  - Economic volatility and global competition demand organizational structures that respond quickly to changing scenarios, which encourages the adoption of agile models and flexible leadership.

These technological and organizational transformations, in turn, have led to a reconfiguration of the corporate culture, which is discussed in more detail in the following sections.

3.3 Current Outlook: Opportunities and Challenges

Digital transformation opens up a range of opportunities, but at the same time poses significant challenges. This duality requires leaders to have a vision capable of **maximizing** the benefits and **mitigating** the risks.

3.3.1 Opportunities

1. Innovation in Products and Services

- AI and automation make it possible to design customer-tailored solutions, fostering customer loyalty and expanding the company's offering.
- Digitalization speeds up cross-functional collaboration, stimulating creativity and the emergence of new business models.

2. Improved Competitiveness in Volatile Markets
   - Real-time data analysis facilitates agile adaptation to changes in demand, optimizing supply chain management.
   - Smaller companies and new entrants can compete with established giants by having access to the same technology infrastructure.
3. Boosting Operational Efficiency
   - Automation of repetitive or manual tasks reduces errors and costs, while human talent focuses on higher-impact tasks.
   - Collaborative tools and virtual platforms eliminate geographical barriers, promoting the proximity of dispersed teams and the optimization of resources.

3.3.2 Challenges

1. Digital Divide and Inclusiveness
   - Not all employees have the same level of digital literacy. Continuous training and professional development plans are essential for technology to be inclusive and non-discriminatory.
   - Cortellazzo et al. (2019) highlight that digitalization generates identity changes in professional profiles, so empathetic leadership is required to accompany teams in these transition processes.
2. Cybersecurity and Technology Governance
   - The more connected the organization, the larger the attack surface. Leaders must advocate for robust security policies and awareness of best practices.
   - The World Economic Forum proposes governance approaches that address the security and sustainability of the digital ecosystem.
3. Resistance to Change and Labor Effects

- Some tasks become obsolete with automation, which generates fear of unemployment and causes internal tensions.
- Clear communications about objectives, support for skills retraining and inclusion of the workforce in decision making help to mitigate fears.

4. Evolving Regulatory Frameworks
    - Artificial intelligence and the massive use of data raise ethical and legal issues (privacy, intellectual property, algorithmic bias).
    - In the absence of uniform regulations on a global scale, organizations must self-regulate and follow international best practices to avoid reputational dilemmas or sanctions.

3.4 The Role of Global Leadership in the Digital Age

In this highly competitive and changing environment, leadership becomes the linchpin of digital transformation. It is not enough to have advanced technologies or large budgets; the culture, motivation and strategy driven by leaders make the difference between success and lagging behind.

1. Vision and Strategy
    - The leader defines the digitalization roadmap, prioritizing projects and ensuring that the technology approach aligns with the organization's overall goals.
    - According to **Deloitte Insights (2020)**, technology leaders with seats on steering committees contribute to smoother adaptation and continuous innovation.
2. Change Management and Communication
    - Explaining not only "what" technology is being adopted, but also "why" and "what for" it is being adopted is essential to counteract resistance to change.
    - A transformational leadership style motivates and accompanies employees, reducing anxiety in the face of automation.
3. Talent and Competency Development

- Digitalization accelerates the obsolescence of some skills, making training and mentoring strategic.
- The leader stimulates curiosity, openness to learning and the emergence of multidisciplinary teams that understand technology and its implications.

4. Ethics and Sustainability
    - Digital transformation can generate exclusion or exacerbate biases if not handled responsibly.
    - Global leaders ensure the implementation of transparency and fairness mechanisms in the use of data, aware of its influence on corporate reputation and customer trust.

3.5 Adjusting the Organizational Culture for Transformation

Organizational culture serves as the matrix that embraces digital transformation. Where there is a culture of collaboration and continuous learning, the adoption of AI and other technological advances is facilitated and enhanced.

1. Experimentation Spaces
    - Controlled pilots and innovation labs provide opportunities to test new tools, encouraging creativity and learning from mistakes.
    - This approach builds team confidence and reduces resistance to broader changes.
2. Technological Inclusivity
    - Training plans adapted to different profiles and career paths.
    - Mentoring programs for personnel with less exposure to emerging technologies, avoiding internal fragmentation of the culture.
3. Values and Ethical Principles
    - Integrate data protection, cybersecurity and social responsibility into the organization's mission and vision.
    - Take into account international guidelines or own regulations that regulate the company's relationship with society and the environment.

4. Participation and Co-creation
    - Multidisciplinary teams, representative of the different areas of the company, generate solutions that are better adjusted to real needs.
    - The involvement of employees in defining priorities and evaluating technologies aligns the strategy with staff commitment.

3.6 Brief Personal Reflection: Convergence of Technology and Finance

In my role as a project manager and consultant specializing in technology and finance, I have found that technology adoption often stumbles more with cultural conceptions or lack of strategic vision than with technical obstacles. In financial automation projects, for example, the use of AI algorithms achieves remarkable efficiencies in estimates and forecasts; however, true success comes only when teams understand the purpose behind the innovation and feel that they are the protagonists of the transformation, not mere executors of a higher mandate.

This experience coincides with the contributions of Cortellazzo et al. (2019), who emphasize the identity dimension of digital change and the importance of leaders fostering trust and a discourse of shared growth. Under this perspective, technology acts as an enabler of competitiveness and resilience, but without inclusive global leadership, its impact is diluted or may even generate a climate of uncertainty that boycotts innovation itself.

3.7 Transition to the Next Chapter

The historical journey and drivers of the Fourth Industrial Revolution, along with the opportunities and challenges identified, illustrate the breadth and complexity of digital transformation. It is not just a matter of deploying technological solutions, but of redesigning culture, processes and competencies. In this scenario, global leadership becomes the linchpin that facilitates or hinders the adoption of innovation.

In Chapter 4, we will explore in more detail how artificial intelligence and automation directly impact strategic decision making, and how these changes link to theoretical models of innovation adoption (Rogers), change management (Lewin) and transformational leadership (Bass). These frameworks will provide a solid perspective to delve into the mechanisms that make technological change sustainable and humane in organizations.

Chapter 4. Artificial Intelligence and Automation: Impacts on Decision Making

Artificial intelligence (AI) and automation are no longer distant concepts but have become an essential part of digital transformation in organizations. These advances not only impact operational efficiency, but also redefine strategic decision making and alter the way leaders lead their teams and manage change. This chapter examines the scope and influence of these technologies on corporate planning, identifies challenges and ethical considerations, and raises concrete implications for leadership in the digital age.

4.1 Definitions and Scope

4.1.1 Artificial Intelligence

Artificial intelligence (AI) can be understood as the ability of computer systems to perform tasks that usually require human intelligence, such as learning, reasoning and decision making (Russell & Norvig, 2016). In the business domain, this definition translates into algorithms capable of:

- **Learning from patterns** in large volumes of data (machine learning).
- **Generate predictive inferences** (sales forecasting models, fraud detection, logistics optimization).
- **Adapt** and refine their results as they receive feedback (deep learning, recommender systems).

4.1.2 Automation

Automation refers to the partial or total replacement of human tasks by technological systems. It is a broader concept that ranges from physical robotization (in assembly lines) to the so-called robotic process automation (RPA) in the field of services (e.g., workflows in accounting or customer service areas). In contrast to AI, which is based on the ability to learn and analyze data, automation is more focused on the systematic execution of repetitive processes.

4.1.3 The Convergence of AI and Automation

With the maturation of the Fourth Industrial Revolution, organizations have begun to combine AI with automation to maximize results, for example:

- Predictive analytics (AI) that detects when an action is required, and then an automated process that executes that action in inventory or logistics systems.
- Intelligent chatbots that not only provide predefined answers, but also analyze natural language to refine their interactions with the user.

This convergence is transforming decision making by enabling an unprecedented level of accuracy and speed, providing leaders with relevant information in real time and streamlining the execution of initiatives.

4.2 Optimization and Strategic Decision-Making

The profound impact that artificial intelligence (AI) and automation have on leaders' strategic decision making in digitized business environments is remarkable. These technologies not only optimize micro-decisions and enable higher volume real-time decisions through algorithm automation, but also improve the quality, efficiency, and accuracy of decision making (Ross & Taylor, 2021). However, they also present significant risks such as over-reliance on algorithms that can lead to the dehumanization of critical decisions, where human judgment remains indispensable (Chui et al., 2018).

In the financial industry, for example, AI is already replacing human tasks in credit analysis and risk management, using advanced algorithms to assess customer creditworthiness faster and with greater accuracy than traditional methods. In manufacturing, automation has transformed assembly lines with robots performing repetitive tasks, increasing efficiency and reducing the incidence of errors (Ross & Taylor, 2021; Chui et al., 2018). Also, the potential of these technologies to transform organizational structure and corporate culture in the long term must be considered. Strategic direction should focus not only on technological efficiency, but also on adaptive and resilient human capital management (Chui et al., 2016).

4.2.1 Transformation of Microdecisions

According to Ross and Taylor (2021), automation can take care of countless routine micro-decisions-for example, approving recurring invoices or identifying credit-risky customers-freeing up leaders' time to focus on strategic decisions. Specifically:

- **Speed**: Machines process large volumes of data uninterruptedly, generating immediate responses.
- **Accuracy**: By reducing the human factor in repetitive tasks, the probability of errors is reduced.
- **Scalability**: Automated systems can handle peak workloads with minimal human intervention, responding to demand without compromising quality.

In practice, this translates into less operational burden for management teams, who can devote their attention to planning new products or services, customer relations and exploring business opportunities.

4.2.2 AI in High-Level Decisions

AI has gone a step further by providing deep learning models that analyze complex correlations in real time. Chui et al. (2018) demonstrate how organizations that integrate AI into their data strategy manage to identify inconspicuous patterns, gaining competitive advantages in budget formulation, market segmentation and early detection of operational problems. Some prominent examples include:

- **Recommendation systems** in streaming or e-commerce platforms, capable of anticipating users' tastes and needs.
- **Demand forecasting models** in retail and manufacturing, which use historical data and external variables (weather, economic activity, etc.) to optimize inventories and operations.
- Financial **trading algorithms**, which operate in fractions of a second by analyzing a multitude of market signals.

These cases show how AI can support leaders in critical decisions, providing a greater degree of certainty than traditional methods, and allowing a more agile reaction to changes in the environment.

4.2.3 The Human Factor: Intuition and Creativity

Although AI provides precision and speed, it cannot replace human intuition, creativity or the long-term vision that characterizes the most important decisions. Therefore, a scenario of complementarity arises:

- The leader interprets the findings of the algorithms and considers contextual variables that are difficult to model in systems.
- Empathy and understanding of social and cultural dynamics remain a human factor domain, essential for negotiations, team motivation and ethical leadership.

In short, AI and automation enhance the ability to analyze and execute, but the strategic direction and ultimate decision responsibility still rests with the vision and judgment of the leader.

4.3 Ethical Challenges and Considerations

4.3.1 Algorithmic Dependence and Dehumanization

A critical risk in placing too much trust in algorithms is **the dehumanization** of decisions that affect people (employees, customers, suppliers). Although models can maximize efficiency, the **ethical dimension** requires evaluation:

- **Algorithmic biases**: If the training data set is not diverse or is contaminated, the system's conclusions could reproduce discrimination or inequity.
- **Lack of flexibility**: A model may ignore human, emotional or contextual variables that are not reflected in the numerical data.

Organizations that opt for extensive automation must always ensure **human control** to review the results and correct for possible biases or unintended consequences.

4.3.2 Privacy and Regulation

The widespread adoption of AI and automation involves **massive data processing**. This brings with it concerns about data privacy and cybersecurity:

- **Sensitive data management**: Banks, hospitals and government agencies handle highly confidential information; misuse or leaks can cause serious damage to user confidence.
- **Legal compliance**: Regulations such as the General Data Protection Regulation (GDPR) in Europe, or others in different regions, require transparency, limits on the storage and correct use of personal data.
- **Shared responsibility**: While leaders must ensure compliance and ethics, IT departments and legal counsel must articulate clear policies to prevent and detect violations.

4.3.3 Labor Impact and Organizational Change

Automation of repetitive tasks may lead to **fear of unemployment** or obsolescence of certain roles. In addition, the new skills demanded by AI change job profiles. This scenario requires:

- **Retraining plan** (reskilling/upskilling): For employees to develop digital competencies and take on higher value-added tasks.
- **Open communication** about the goals of automation: Explaining that technology frees up time for more creative or analytical tasks reduces resistance to change.
- **Reconfiguring structures**: With AI and automation, some traditional hierarchies are becoming more flexible, giving rise to agile, multidisciplinary teams that integrate data science experts, automation engineers and leaders with cross-functional vision.

4.4 Implications for Leadership

4.4.1 Digital Leadership and Knowledge Management

In an environment where AI and automation take over much of the operational tasks, the leader must excel at:

1. **Information Curation**: Knowing what data and AI results really contribute to the strategic objective.
2. **Knowledge Dissemination**: Ensure that insights generated by algorithms do not remain in silos, but are integrated into the organization's culture and decision-making routines.
3. **Promoting Transparency**: Explain how certain AI models work (to the extent possible) and why they are adopted so that the team understands the benefits and limits.

4.4.2 New Focus on Soft Skills

Increasing automation places greater emphasis on human skills that cannot be programmed:

- **Emotional Intelligence and Empathy**: Essential for navigating staff fears and building trust in change processes.
- **Effective Communication**: Leaders must translate AI findings into clear action plans aligned with the organization's culture.
- **Critical Thinking**: Review and question the results of the models, avoiding a blind dependence on algorithms.

4.4.3 Strategic Decisions with Human-Technological Balance

The role of the leader evolves into a **mediator** that combines AI precision with human sensitivity and long-term perspective. As several studies argue (Ross & Taylor, 2021; Chui et al., 2018), the greatest impact occurs when:

- **AI** and automation are focused on boosting efficiency and innovation.
- **Leadership** is responsible for guiding cultural adaptation, training staff and ensuring ethics and sustainability.

4.5 Chapter reflections

The incorporation of AI and automation represents a radical change in the way decisions are made within organizations. On the one hand, it brings speed and accuracy; on the other, it raises questions about ethics, the role of human talent and long-term sustainability. Leaders must therefore adopt a critical and balanced stance: harness the power of technology while preserving the human essence that brings meaning, empathy and responsibility to every choice.

In the next chapter, we will explore how these changes dovetail with the theoretical frameworks already presented - Innovation Adoption Model (Rogers), Organizational Change Theory (Lewin) and Transformational Leadership (Bass) - to understand why resistance to change, culture management and team motivation are so critical to the success of digital transformation. As AI and automation consolidate their presence in decision making, leadership will increasingly need skills to navigate with empathy, communicate with transparency and ensure sustainability in an environment dominated by relentless innovation.

This reaffirms that global leadership in the digital era cannot be limited to the passive adoption of technological tools, but must orchestrate the convergence between automated processes and human talent, fostering both agility and the cohesion and commitment of all the actors involved.

Chapter 5. Digital Leadership Skills and the Ability to Adapt.

Digital transformation requires leaders to have a new set of competencies and a mindset open to continuous change. If in the past it was enough to exercise authority and coordination, in the era of artificial intelligence (AI) and automation it is required to combine technological vision with empathy, adaptability and constant learning. This chapter delves into the digital leadership skills that arise from the integration of technology into organizational processes, emphasizing soft skills, change management and the need to promote agile cultures. Both conceptual elements and practical examples will be addressed to illustrate how leaders can cultivate adaptability in an ever-changing business environment.

5.1 Digital Leadership: Towards a Multidisciplinary Approach

Digital leadership is not limited to possessing specific technical competencies, but implies understanding technology as a catalyst for cultural processes and profound transformations in the way of working. According to Cortellazzo et al. (2019), the leader who spearheads digital innovation must be able to:

1. **Communicate the vision** of how technology drives growth and sustainability.
2. **Articulate adoption processes** (following, for example, Rogers' Model on the diffusion of innovations).
3. **Connecting the implications of change** with the well-being and professional development of the teams.

This leadership requires the ability to **mobilize resources** - both financial and human - to make digitization an inclusive process consistent with strategic objectives.

5.2 Key Digital Leadership Competencies

5.2.1 Technology Vision and Curiosity

To guide organizations in the digital age, leaders need a technological vision that goes beyond the merely operational. It is not essential to master all programming languages or algorithms, but it is essential to have the curiosity to:

- **Explore** trends in AI, automation, big data or cloud computing.
- **Evaluate** the potential of new solutions and how they align with corporate strategy.
- **Anticipate** the disruptions that these technologies may cause in the market or in the business model.

Curiosity drives the leader to **question**, **learn** and **adapt**, ensuring a vision of the future that prevents organizational stagnation.

5.2.2 Digital Literacy and Competencies

While not all managers need to be programming experts, mastery of certain digital skills becomes a differentiating factor:

- **Data literacy**: Understand basic analytics and statistical concepts to interpret performance reports and dashboards.
- **Collaborative tools**: Learn about digital platforms and ecosystems (Teams, Slack, SharePoint, etc.) that facilitate communication and information exchange.
- **Security and privacy management**: Be aware of the main cyber risks and data protection regulations to promote information integrity.

These competencies generate greater autonomy and criteria for dialogue with technical teams, avoiding total dependence on external advisors or middle management. In today's dynamic business environment, marked by rapid technological advances, digital leadership skills and the ability to adapt are not only desirable, but essential. Leaders who master these competencies are better equipped to implement and maximize emerging technologies, directly influencing organizational productivity and efficiency.

According to Chamorro-Premuzic (2021), effective leaders in managing digital technologies can increase the productivity of their teams by up to 24% and operational efficiency by 30%. However, it is vital that these advances are not pursued at the expense of the human aspects of the organization. A purely technical approach could neglect critical staff needs, such as wellbeing and continuous professional development, which are essential for long-term sustainability.

5.2.3 Strategic Thinking and Adaptability

In a highly volatile environment, strategic thinking gains relevance by linking technological adoption with long-term objectives. To this end, adaptability emerges as a core competency:

- Rapidly **reconfigure priorities** in the face of new digital opportunities or threats.
- **Learn from** implementation **failures** and reuse that knowledge to refine the strategy.
- **Apply agile methodologies** that fragment projects into shorter phases, with continuous feedback (DevOps, Scrum, Lean, etc.).

Adaptability is not reduced to the leader's individual action; it also involves his or her ability to promote agility throughout the organization. Klus and Müller (2021) argue that keeping up with the latest technological trends is crucial for contemporary leaders. However, focusing exclusively on technology can lead to dehumanized management, where technological efficiency overshadows the importance of human connection and ethical leadership. Therefore, the leader must balance technical competence with strong interpersonal skills.

Digital agility must be cultivated as a core competency in modern leadership, as examined by Sambamurthy et al. (2003). They propose that flexibility and adaptability are as crucial as the adoption of the technology itself. Rigid and non-adaptive leadership can result in a lack of responsiveness to constantly evolving market dynamics, putting the organization's competitiveness at risk. Likewise, Khan (2016) highlights the importance of developing adaptive skills for effective collaboration with machines, increasing organizational competitiveness by 25%.

5.2.4 Virtual Communication and Remote Team Leadership

The increase in remote work and virtual collaboration requires specific communication skills:

- **Clarity and empathy**: Digital interaction can generate misunderstandings due to the absence of non-verbal cues. The leader must ensure that objectives, timelines and expectations are unambiguously understood.
- **Personal connection**: Informal exchange spaces (virtual coffees, check-ins) can be crucial to reinforce team cohesion and motivation.
- **Coordination of time zones**: Globalization and geographic dispersion imply managing diverse calendars and respecting multiculturalism.

Here, soft skills - emotional intelligence, active listening, constructive feedback - are intertwined with digital literacy and the initiative to cultivate a collaborative environment at a distance.

5.2.5 Emotional Intelligence and Empathy

Artificial intelligence and automation create uncertainty in the workforce. Some people fear the obsolescence of their roles; others may feel overwhelmed by the need to learn new technologies. The digital leader must:

- **Practice empathy** to identify concerns and promote an open dialogue about the change process.
- **Provide psychological safety**: Allowing employees to explore solutions and make controlled mistakes fosters creativity and reduces resistance to change.
- **Balancing attention** between results and well-being, recognizing that the human dimension is essential to the success of any technological initiative.

The intersection of human and technological skills constitutes a central axis for leadership effectiveness in the digital era. Avolio and Kahai (2003) emphasize that the ability to effectively manage one's own and others' emotions is crucial to facilitate adaptation to new technologies. Ignoring these human aspects can result in teams that are less committed and reluctant to embrace change, which undermines the organization's digital evolution

5.2.6 Ethics and Social Responsibility

Modern organizations cannot ignore the ethical and social implications of their technological actions. The digital leader must:

- **Ensure fairness** in the application of algorithms, minimizing bias and monitoring the use of sensitive data.
- **Promote inclusion** and diversity, ensuring that digitalization benefits all groups and does not widen existing gaps.
- **Promote sustainability** in technological initiatives, evaluating the environmental and social footprint of the solutions adopted (World Economic Forum, 2021).

Ultimately, the credibility of digital leadership depends largely on how these ethical commitments are made. Sheninger (2019) warns that adopting technologies without a strategic and ethical approach can lead to technology initiatives that, while flashy on paper, fail to generate sustainable value if they do not reinforce organizational culture and promote a sense of belonging and purpose.

5.3 Adaptive Capacity and Organizational Resiliency

The capacity to adapt implies both the agility to react to immediate changes and the resilience to withstand prolonged crises. This organizational resilience is strengthened by the strength of leadership and the trust that the leader generates in his team.

5.3.1 The Transformational Leadership Role

Bernard Bass' Transformational Leadership Model stresses that leaders who inspire, motivate and give individualized attention foster a culture of positive change (Bass & Riggio, 2006). In the digital realm, this translates into:

- **Inspirational motivation**: Set clear and ambitious goals for technology adoption, communicating in an attractive way the benefit for the organization and for employees.

- **Intellectual stimulation**: Invite teams to question obsolete processes and propose innovative solutions.
- **Individualized consideration**: Recognize the different needs and capabilities of each member, customizing training or support in the digital transition.

In this sense, Bass and Riggio (2006) also point out that transformational leaders can foster a 40% increase in innovation by creating a climate that favors adaptation and experimentation with new technologies. This type of leadership not only drives technological advancement, but also promotes a culture of continuous learning, where collaboration and receptivity to change become essential pillars.

This blend of leadership and empathy creates an environment that fosters adaptability and resilience to the challenges of transformation.

5.3.2 Continuous Learning and Iteration

Adaptability requires leadership to promote a continuous learning approach:

- **Regular training plans**: Refresher workshops, online courses and internal trainings that address from specific digital tools to new management methodologies.
- **Agile feedback**: Periodic review sessions, in which technological advances are reviewed, deviations are corrected and objectives are adjusted.
- **Constant iteration**: Implementing digital solutions often requires successive improvements. Adopting an iterative approach minimizes the risk of major failures and maintains team motivation.

5.3.3 Cases and Brief Examples

- **Case of a retail company**: After implementing an AI system to suggest products to online customers, the marketing team detected the need to reinforce analytical skills. A mentoring plan and internal certifications were organized, accelerating the learning curve and reducing dependence on external consultants.

- **Financial services company**: Integrated agile methodologies into process automation adoption, enabling controlled experiments with RPA. After each sprint, the team discussed lessons learned, adjusted objectives and designed more sophisticated prototypes.

In both scenarios, adaptability was enhanced by a leadership that combined vision, accompaniment and focus on the development of digital competencies.

5.4 Personal Reflection: A Look at Digital Leadership in Practice

In my work as a Project Manager and expert in technology and finance, I have witnessed the importance of flexibility and empathy in the adoption of digital innovations. On the one hand, managing projects involving AI and automation requires a certain technological literacy to understand the real opportunities and limitations of each tool. On the other hand, the human component is decisive: I have found that teams are more open when they feel that the leader values their learning and their concerns, rather than imposing tools without a clear narrative.

Aligned with Rogers' Model on the diffusion of innovations, the leader's willingness to listen and "evangelize" the usefulness of technology - with tangible evidence and training plans - drives adoption. In turn, Bass' transformational vision nurtures a sense of belonging and intrinsic motivation, key factors in keeping the organization on track in the midst of rapid change.

5.5 Chapter reflections

Digital leadership skills and adaptability become imperative in a business environment characterized by the confluence of AI, automation and profound organizational changes. Core technical competence, combined with effective communication, empathy and a transformational approach, constitutes the ideal profile for leading teams in the era of the Fourth Industrial Revolution.

A pragmatic perspective is provided by Asatiani et al. (2023), who point out that proper implementation of robotic process automation can reduce operating costs by up to 50%. However, the transition requires leadership that not only focuses on short-term gains, but also considers the human and organizational impact of the technology, ensuring adoption that benefits the entire organization.

Likewise, leaders who manage to balance human and technological capabilities can increase organizational competitiveness by up to 25% (Khan, 2016). This balance is essential for the digital era, where creativity and empathy remain irreplaceable by technology.

Similarly, the seamless integration between technological skills and human factors becomes central to truly effective leadership in this age of disruption. Sambamurthy et al. (2003) insist that agility and adaptability are as crucial as adoption the technology itself, reinforcing the idea that rigid leadership can fall behind in the face of rapid market dynamics. Finally, Sheninger (2019) reminds that a technology adoption without a strategic and ethical approach can result in superficial initiatives that do not generate sustainable value or connect with employees' sense of belonging and purpose.

In the next chapter, we will delve into the strategies and tools that leaders can employ to put these skills into practice, ensuring that digital transformation is not just talk, but a sustainable and successful reality in terms of growth, innovation and cohesion of human talent.

Chapter 6. Resistance to Change and Organizational Transformation

Digital transformation involves not only adopting emerging technologies, but also revising the structures, processes and work dynamics that have governed the organization. This reformulation entails substantial changes that can raise resistance at all levels, from operational employees to senior management. This chapter explores the factors that generate resistance to change, analyzes Kurt Lewin's Theory of Organizational Change as a framework, and illustrates practical strategies to facilitate the transition to digitized environments. It also presents examples of organizations that have faced significant challenges in implementing automation, artificial intelligence and cultural restructuring processes.

6.1 Nature of Resistance to Change in the Digital Age

The implementation of digital transformation, while an urgent business need, presents challenges and resistance that require astute and thoughtful leadership to overcome. This analysis not only explores strategies for change, but also underscores the need for a critical approach in assessing both the potential benefits and risks associated with the digitization process.

6.1.1 Concept of Resistance

Resistance to change manifests itself as a set of attitudes or behaviors that hinder the adoption of new technologies, work methods or organizational structures. In the context of digital transformation, it can present itself in many different ways:

- **Fear of obsolescence**: Employees fear that automation or AI will replace their jobs, or that they will not have the skills to adapt to new requirements.
- **Distrust in technology**: Lack of information or previous negative experiences generate skepticism about the real value of digital tools.
- **Cultural inertia**: Ingrained processes and a "traditional" corporate culture favor the continuity of obsolete practices and hinder the implementation of agile methodologies.

- **Short-term vision**: Managers fear the cost and complexity of change, postponing technological investment or relegating it to pilot projects without continuity.

6.1.2 Key Factors Affecting Resilience

1. Lack of communication
   - When management does not clearly explain the objectives of the change, its benefits and the impact it will have on the employees' routine, it generates uncertainty and rumors that fuel resistance.
2. Lack of training
   - If the organization does not provide adequate training, staff may feel that they do not have the necessary skills for the new digital era, generating anxiety and rejection.
3. Changes in Power and Status
   - Digitization may redistribute functions, visibility or resources within the company, so those who see their "turf" threatened may consciously oppose modernization projects.
4. Previous experiences
   - Poorly managed change initiatives in the past (costly technological failures, introduction of unfriendly systems) generate skepticism towards any new transformation proposal.

6.2 Kurt Lewin's Theory of Organizational Change

Kurt Lewin's Theory of Change is based on a three-phase process - Unfreezing, Changing and Refreezing - which, despite its historical nature , is still extremely useful for understanding the dynamics of digital transformation.

Kurt Lewin's change model describes organizational change as a three-phase process: unfreezing, changing and refreezing. In the unfreezing phase, the organization prepares for change by challenging existing norms. The change phase introduces new practices, requiring effective leadership to motivate staff. Finally, refreezing solidifies these changes as the new norm. Without proper management of these phases, organizations can experience significant resistance.

However, although Kurt Lewin's model provides a theoretical basis for addressing organizational change, it is crucial to recognize that this model, while robust, may oversimplify the complexity of digital transformation. The unfreezing phase, in particular, requires more than mere preparation for change; it demands constant challenging of perceptions and continual reassessment of strategies as new technologies emerge.

6.2.1 Thawing

It is about challenging the status quo and getting the organization to recognize the need for change. In terms of digital transformation, this includes:

- **Awareness creation**: Present data and results that demonstrate the obsolescence of current tools or the risk of losing competitiveness if innovation is not implemented.
- **Identifying sponsors**: Having influential leaders to champion the change and explain its importance.
- **Awareness and empathy**: Addressing staff concerns, understanding fears and explaining tangible benefits (opportunities for professional growth, improved quality of work).

6.2.2 Change

In this phase, the organization introduces new processes, technologies or behaviors. It is the most complex stage, as it requires a redesign of the culture and a real test of adaptability.

- **Implementation of digital solutions**: AI, RPA, collaborative platforms, etc.
- **Training and support**: Reskilling and upskilling programs so that human talent is empowered and uses the tools effectively.
- **Piloting methodologies**: Conduct controlled tests (pilots) and continuous learning sprints that reduce risks and allow adjusting errors in the short term.

6.2.3 Refreezing

It consists of consolidating the new practices so that they become the norm and take root in the organizational culture. It involves:

- **Positive reinforcement**: Recognize achievements and celebrate the benefits of technology adoption, reinforcing employee confidence.
- **Adjustment of structures**: Update manuals, protocols and organizational charts to reflect the new ways of working.
- **Continuous monitoring**: Maintain indicators that ensure the permanence of changes and constant improvement, avoiding the temptation to return to previous practices.

6.3 Strategies for Overcoming Resistance to Change

Ross and Beath (2002) stress the importance of IT governance, but it should be criticized that governance often focuses too much on the technical aspects, neglecting the human and cultural needs that are equally vital to a successful transition. Technology should not dictate the direction; rather, it should be a tool in the hands of visionary leaders who understand its impact on people and organizational processes.

Powell et al. (2017) argue about tailoring implementation strategies to the specific context of each organization. This perspective is valuable, but the criticism also emerges that companies sometimes adopt technologies more out of competitive pressure than genuine needs assessment. This "technology leapfrogging" can result in initiatives that are superficial and not deeply integrated into business operations.

6.3.1 Effective Communication and Project Narrative

Proactive and transparent communication reduces uncertainty and rumor. Some key tactics include:

- **Vision Statement**: Explain the "why" of the transformation, connecting the technological objectives with the company's mission and competitiveness.

- **Internal testimonials**: Leaders or teams that have experienced improvements thanks to digitization share their experience, motivating others to join them.
- **Feedback channels**: Surveys, virtual forums or Q&A sessions to clarify doubts and listen to concerns in real time.

6.3.2 Training and Accompaniment

Ongoing training dispels the fear of obsolescence and reinforces the sense of control over change:

- **Reskilling plans**: Focused on digital competencies (data analytics, cybersecurity, agile methodologies).
- **Cross-mentoring**: Employees with technological skills support those who require a closer accompaniment.
- **Learning by Doing**: Practical learning through concrete projects and collaborative workshops that accelerate the actual adoption of the tools.

6.3.3 Participation and Co-creation

When employees feel they are an active part of the change, resistance decreases. Leadership can:

- **Form innovation groups**: Bring together people from different departments to contribute ideas and design prototypes for improvement.
- **Gamification of the process**: Implement playful dynamics that reward achievement and learning in the digital transformation process.
- **Decentralize decisions**: Give greater autonomy to self-managed teams that can adapt technology to their needs and realities.

6.3.4 Transformational Leadership and Emotional Support

Individualized consideration and inspirational motivation, put forward by Bass (1985), are essential to emotionally accompany people. The transformational leader:

- **Actively listens** to concerns and is responsive, conveying confidence and clarifying expectations.

- **Reinforces the** team'**s self-esteem** by highlighting the opportunity for growth rather than the fear of replacement.
- **Empower** employees to discover new skills, celebrate breakthroughs and integrate transformation into their professional development.

Schwartz and McCarthy (2007) and Kane et al. (2015) stress the need for digital competence in leadership and the importance of addressing staff concerns about job security. However, leaders must go beyond mere training and assurance by actively involving staff in the design and execution of digital transformation. This collaborative approach can reduce resistance by facilitating a sense of ownership and control over the change process.

6.4 Examples of Organizations Facing Digital Transition Challenges

Success stories like Ford and General Electric are instructive, but it is also essential to recognize that what works for a global corporation may not be applicable in a smaller company or a different industry. Success in digital transformation is not simply a matter of following examples, but of intelligently adapting lessons learned to the unique realities of each organization.

6.4.1 Manufacturing Company with Obsolete Systems

A traditional organization in the industrial sector found that its production control and recording systems were no longer keeping up with global demand. When faced with a proposal to migrate to an IoT platform with real-time dashboards, there was significant resistance from veteran technicians who feared they would be left behind.

- Strategy:
    - A gradual training plan was designed for personnel with less digital familiarity.
    - Mixed committees (senior technicians and young professionals) were created to implement the sensors and validate the results.
    - Management encouraged collaboration by inviting the more experienced to pass on process knowledge to the newcomers.

- Result:
  - Within months, resistance diminished as workers found that the new system not only increased efficiency, but also improved accuracy in maintenance scheduling and quality monitoring.

6.4.2 Financial Services Corporation and the Migration to IA

A financial services company decided to incorporate AI models for credit risk analysis and fraud detection. However, some analysts and managers questioned the reliability of the algorithms, arguing that financial decisions required experienced human judgment.

- Strategy:
  - AI models were integrated gradually, first as recommendation tools that supported the analyst without replacing him.
  - The basic operation of the algorithms was made transparent, showing examples of successful predictions and false positives/false negatives.
  - Lewin's theory was applied: unfreezing fears through lectures and workshops, changing credit evaluation practices with continuous accompaniment and refreezing by establishing a final human review protocol.
- Result:
  - Progressive adoption and the combination of human expertise with the accuracy of the algorithm increased the speed of response to customers and reduced fraud levels by 20%.
  - Staff felt that their judgment was still valued, minimizing the perception of "dehumanization" in financial decisions.

6.5 Personal Reflection: Change as a Constant

In my experience as a Project Manager specialized in technology and finance, I have found that resistance to change appears even in organizations that claim to be "innovative". The key lies in communicating, accompanying and democratizing the transformation process. When people understand why new tools are being introduced - and how this contributes to their professional development - the reaction usually shifts from opposition to cautious enthusiasm, and from there to full integration.

I have seen teams transform completely when the leader transmits confidence in the training of their employees and gives space for each one to take a leading role in digitization. In the end, any resistance dissipates when the culture is aligned with the vision of a future that is more competitive, but also more human and collaborative.

6.6 Chapter reflections

Resistance to change is a natural phenomenon in any transformation process, and digital transformation is no exception. Through Lewin's Model, the importance of raising awareness (unfreeze), implementing with accompaniment and training (change) and consolidating new habits (refreeze) becomes evident so that the organization assumes a culture open to innovation. While digital transformation is essential for business competitiveness, its successful implementation requires a delicate and thoughtful balance between technology and humanity. Leadership that values both technological innovation and cultural and human integrity is essential to transform challenges into opportunities and guide organizations towards a sustainable digital future.

In the next chapter, we will delve into how people and technology can be synergistically integrated, exploring the concept of "hybrid leadership" and the relevance of balancing technical competencies with emotional intelligence, ethics and long-term vision. As digital transformation advances, the human component becomes essential to ensure that the solutions implemented generate sustainable value and promote a culture of constant collaboration.

Chapter 7. Integrating People and Technology: Hybrid Leadership

Digital transformation is not limited to the introduction of technological tools; it requires a comprehensive reconception of the way in which people interact with technology and, above all, the role that people play in automated or artificial intelligence (AI)-powered environments. Hybrid leadership emerges precisely as the answer to this challenge: leading by combining human competencies - empathy, creativity, communication - with the technical and cognitive skills of the digital era. This chapter explores how to integrate people and technology in a harmonious way, addressing the benefits and tensions generated by bringing these two worlds together in the organizational culture.

7.1 The Concept of Hybrid Leadership

Hybrid leadership involves effectively combining human attributes with technological capabilities to orchestrate the dynamics of a team and/or an organization in a digital environment. Although the previous chapters emphasized the importance of emotional intelligence, change management and digital competencies, this time we delve deeper into how this union transcends the sum of these competencies:

1. **Human-technology synergy**: It is not a matter of technology replacing human judgment, nor of people ignoring the advantages of automation. The hybrid leader articulates both elements to maximize productivity and organizational well-being.
2. **Co-creation and experimentation**: Hybrid leadership encourages openness to experimentation with digital tools and new processes, while protecting space for reflection, collective learning and continuous innovation.
3. **Strategic vision and cultural sensitivity**: The leader understands the cultural and psychological impact of AI and automation, and proactively manages team resistance and expectations.

In essence, hybrid leadership is based on a deep understanding of human strengths (intuition, creativity, empathy) and technological benefits (efficiency, accuracy, massive data management).

7.2 People and Technology: New Ways of Working

7.2.1 Towards Hybrid Collaborative Teams

The increasing automation of tasks drives the creation of mixed teams composed of people and AI systems or collaborative robots (cobots). In this context:

- Repetitive and high-precision tasks are delegated to technology.
- **Analytical, supervisory and decision-making tasks** (especially those requiring human sensitivity) are left to trained professionals.
- **Coordination** becomes a key factor: the dynamic allocation of activities, the management of shared data and the ability to react quickly to contingencies require leaders who manage logistics, communication and motivation simultaneously.

7.2.2 Redefinition of Roles and Responsibilities

The introduction of AI and automation blurs traditional job boundaries. Some roles become obsolete, while others are transformed or new ones emerge:

- **Hybrid systems specialists**: People who master both the human side (communication, training) and the technical side (parameterization and continuous improvement of algorithms).
- **Learning facilitators**: Responsible for designing reskilling and upskilling plans, aligned with the technological evolution of the organization.
- **Data analysts with a human perspective**: Combine advanced analytics with contextual reading of information, avoiding the overreliance on algorithms and complementing the results with cultural or situational criteria.

This new landscape blurs rigid hierarchies and favors a multidisciplinary model where adaptability takes precedence over narrow specialization.

7.3 Essential Human Competencies in Hybrid Leadership

7.3.1 Emotional Intelligence and Empathy

When processes are automated and AI systems are introduced, uncertainty about the future of jobs and responsibilities is common. Therefore, hybrid leadership involves a high degree of empathy for:

- **Actively** listen to the concerns of employees.
- **To design support interventions** that reinforce psychological safety in the teams.
- **To value the human contribution** in processes that, apparently, could be completely digitized.

In line with Bass' (1985) vision, the leader transforms the fear of change into an opportunity for growth and learning, promoting a culture of innovation with a human sense.

7.3.2 Critical and Ethical Thinking

Technology provides powerful tools, but a hybrid leader must know how to question AI results, identify potential biases in algorithms, and consider the ethical implications of their decisions. Examples:

- **Evaluate fairness** in automated decisions (hiring, credit, diagnostics) to avoid implicit discrimination due to biased data.
- **Protect the privacy** of customers and employees, ensuring responsible collection and use of information, aligned with corporate standards and values.
- **Adopt a sustainability approach**: from selecting cleaner technologies to looking at the carbon footprint of digital infrastructure.

This ethical approach reinforces the organization's and the environment's **confidence** in hybrid management, demonstrating consistency between what is done and the principles that are proclaimed.

7.3.3 Adaptability and Continuous Learning

In environments where technology changes rapidly, the hybrid leader must demonstrate adaptability and foster that same ability in his or her team:

- **Flexible training plans**: Incorporate agile methodologies to train personnel in new tools without abruptly interrupting operations.
- **Iterations and retrospectives**: When adopting digital solutions, establish a testing and adjustment cycle that allows for continuous improvement and open feedback.
- **Constant updating**: Keeping up to date with trends in AI, robotics and emerging solutions to anticipate potential disruptions or opportunities in the market.

7.4 Strategies for Optimal Integration of People and Technology

7.4.1 Culture of Collaboration and Co-creation

Working together between human teams and intelligent systems flourishes in an environment where collaboration is encouraged. To achieve this:

- **Form mixed committees** that involve technical profiles, business experts and collaborators from different areas, co-creating digital solutions that respond to specific needs.
- **Promote early participation** in the selection or design of technologies. The earlier end users see and test a tool, the more feedback is generated and the faster resistance is eliminated.

7.4.2 Design Thinking and Pilot Experiences

Adopting a design thinking approach translates into a people-centric vision, even when the goal is to increase automation:

- **Understand the real needs** of internal users or end customers before programming a system or deploying a collaborative robot.
- **Validate prototypes in early stages** (pilot projects) minimizing resistance and investment of time and resources.

- **Iterative feedback**: pain points are quickly detected, adjusting the implementation.

7.4.3 Change Management and Management Sponsorship

While hybrid leadership emphasizes the role of the person articulating technology and culture, senior management must support with:

- **Sufficient budget** to invest in infrastructure, software licenses, training and hiring of experts.
- **Narrative support**: Speeches and communications that legitimize the importance of innovation, reducing the gap between the top and the operational teams.
- **Recognition of learning**: Reward employees and middle leaders who demonstrate initiative and competence in the integration of technology and human labor.

7.5 Challenges and Tensions in Hybrid Leadership

Even with a good strategy, tensions arise from the hybridization of people and technology:

1. **Potential job displacement**: Fear of losing jobs accelerates or exacerbates resistance. The hybrid leader must communicate with clearly how staff realignment or specialization is an integral part of the roadmap.
2. **Biases in algorithms**: Lack of diversity in data or model training can lead to unfair decisions. The leader requires supervision and correction mechanisms.
3. **Information overload**: AI and automation systems can generate large volumes of data. Lack of prioritization and analysis can lead to decision paralysis by analysis.
4. **Lack of cultural cohesion**: Remote or geographically dispersed teams with varied digital competencies can suffer from disconnection and loss of shared identity.

Hybrid leadership not only detects these risks, but designs strategies to anticipate or mitigate them, ensuring that the transformation project flows with a clearly communicated purpose and talent management that enhances human qualities.

7.6 Chapter reflections

Adopting hybrid leadership reveals that the true value in digital transformation lies not just in automating processes or implementing artificial intelligence systems, but in how the strengths of the human factor are married with the advantages of technology. The leader who achieves this integration understands that creativity and empathy remain irreplaceable, even when algorithms provide accurate information or execute repetitive tasks efficiently.

Rather than viewing people and machines as polar opposites, hybrid leadership seeks to cultivate a relationship of synergy: technology bolsters productivity and amplifies analytical capacity, while human teams bring the ethical judgment, adaptability and emotional intelligence needed to navigate uncertain and changing environments. From this perspective, automation should not be feared as a threat, but embraced as an opportunity to unleash the creative and strategic potential of human talent.

The following chapter will present concrete strategies and tools for digital transformation, showing how leaders can structure action plans that materialize this human-technology collaboration in a successful and sustainable way, ensuring the competitiveness of the company and, at the same time, the integral growth of its employees.

Chapter 8. Strategies and Tools for Digital Transformation

Digital transformation goes beyond the mere adoption of technology. It involves structuring a coherent plan - supported by clear methodologies, tools and metrics - that enables organizations to plan, execute and evaluate their initiatives consistently and in alignment with their business objectives. This chapter presents key strategies and tools to guide digital transformation, from the design phase to the results monitoring phase, covering technology selection, innovation governance and the measurement of financial and cultural impacts.

8.1 Roadmap for Technology Adoption

The creation of a roadmap is essential to structure the organization's various technology initiatives. An effective roadmap includes:

1. **Definition of objectives**
    - Clarify transformation goals: Are you looking to increase operational efficiency? Optimize customer experience? Explore new digital business models?
    - Connect the technological vision with the long-term corporate strategy.
2. **Project prioritization**
    - Analyze the **feasibility** (technical, economic, cultural) of each project and the **value** it brings.
    - Stagger initiatives in phases, starting with those that generate "quick wins" and establish confidence in the transition.
3. **Identification of resources and competencies**
    - Determine the **internal skills** required (data analytics, cybersecurity, agile methodologies) and the **external capabilities** that could complement them (consulting firms, suppliers, strategic alliances).

   - Establish an **estimated budget** and a financing plan (equity, loans, venture capital, etc.).

4. **Timeline and lines of responsibility**
   - Delimit milestones and deadlines, assigning leaders or teams responsible for each project.
   - Ensure that monitoring and periodic review mechanisms are in place to adapt the roadmap if changes in the environment arise.

This plan allows the organization to coordinate efforts, avoid overlaps and manage resources optimally. It also serves as an **internal communication tool** for all employees to understand the "big picture" of the transformation.

8.2 Integration of New Technologies in Daily Operations

Even with a good roadmap, the effective integration of digital solutions into business-as-usual processes can present challenges. To overcome the friction between old routines and new tools, it is advisable to:

8.2.1 Evaluation and Redesign of Processes

- **Identify bottlenecks** and areas with duplicated or manual tasks that are amenable to automation.
- **Involve the teams** that execute these processes, gathering their opinion on improvement points and the real feasibility of the digitization proposals.
- **Simplify before automating**: Review whether the process can be optimized or partially eliminated before implementing a costly system.

8.2.2 Change Management and Focused Training

- **To train "change agents"** in each unit, responsible for promoting and accompanying the adoption of digital tools within their respective teams.

- **Design training** oriented to specific operational needs: it is not enough to teach theory; it is essential to show how technology speeds up or improves daily work.
- **Offer continuous support** (help desk, internal forums, online manuals) to resolve doubts and encourage the use of new solutions.

8.2.3 Progressive Adoption and Pilots

- **Test in pilot areas or teams** before a massive launch of the solution, validating usability and acceptance by personnel.
- **Gather feedback** and improve the tool or implementation method before scaling up.
- **Celebrate** early **achievements**, disseminating positive results to encourage the rest of the organization to get on board with the change.

8.3 Innovation Governance and Management Tools

Digitization requires strong governance to ensure alignment with strategy, transparency in resource allocation and risk mitigation. Some relevant tools and practices include:

8.3.1 Innovation Committees and Digital Transformation Offices

- **Innovation Committees**: Bring together leaders from different areas, including finance, technology, HR and business units, to prioritize initiatives and monitor progress.
- **Digital Transformation Office (DTO)**: It works as a facilitator, with clear mandates to promote digitalization projects, to accompany the teams in the adoption and evaluate the return on investment (ROI).

8.3.2 IT Governance Models and Agile Methodologies

- **IT Governance Framework (COBIT, ITIL)**: Provides guidelines for the control and optimization of technology services and processes, ensuring quality and security in the delivery of value.

- **Agile Methodologies (Scrum, Kanban)**: Allow to iterate quickly, launch early versions of digitized products or processes and adjust the project based on real user feedback.

8.3.3 Risk Management and Cybersecurity

- **Digital risk analysis**: Identify cybersecurity threats, compliance issues and operational vulnerabilities associated with digitization.
- Information security **policies and protocols**: Define specific measures (encryption, multifactor authentication, contingency plans) to protect critical data and infrastructure.
- **Awareness**: Conduct awareness campaigns so that all employees are aware of their responsibilities in terms of security and confidentiality.

8.4 Metrics and Success Indicators

For digital transformation to be sustainable, it is vital to measure the impact and progress of each initiative. These metrics may include:

8.4.1 Process Indicators

- **Adoption rate of** new technology (percentage of employees who use it regularly).
- **Average response time** in digital services, compared to the previous situation.
- **Reduction of errors** in automated processes.

8.4.2 Financial and Return on Investment (ROI) Indicators

- **ROI of projects** (benefit/cost ratio, payback period, IRR).
- **Increase in sales or revenue** attributable to advanced analytics or new e-commerce platforms.
- **Operational savings** by automating repetitive tasks.

8.4.3 Cultural and Organizational Climate Indicators

- **Employee satisfaction indexes** (surveys, interviews) to measure the perception of digitalization and satisfaction with the training received.
- **Level of commitment** to innovation (number of ideas proposed, participation in pilot projects).
- **Lower resistance**: For example, a decrease in formal complaints or an increase in attendance at training sessions.

Once this data is obtained, leaders can readjust the roadmap, identify obstacles and celebrate achievements, increasing the credibility of the transformation.

8.5 Chapter reflections

The strategies and tools presented in this chapter outline a practical framework for approaching digital transformation in a planned and structured manner. From developing a clear and feasible roadmap to implementing governance policies and measuring the actual impact of initiatives, each step contributes to making digitization a sustainable project aligned with the organization's objectives.

The key lies in combining the agility of innovation methodologies with the rigor of results monitoring. In this way, technology goes from being an abstract promise to a tangible driver of growth and improvement in the experience of employees and customers. However, it is important to remember that digitization does not happen in a vacuum: its success depends on the culture, leadership and trust of those involved in it.

In the following chapter, we will analyze case studies and lessons learned from organizations that have successfully -or stumbled- along the path of digital transformation, offering concrete examples that complement and enrich the strategies outlined here. This will reinforce the vision that the adoption of emerging technologies is a continuous journey, in which informed decisions and the integration of all human talent are the fundamental pillars to achieve a real and lasting transformation.

Chapter 9. Case Studies and Lessons Learned

Digital transformation manifests itself differently in each organization, depending on factors such as culture, business strategy and industry characteristics. However, technology adoption processes often converge on common challenges and lessons learned, which can serve as a reference for leaders seeking to implement similar changes. This chapter presents case studies - some based on real examples and others hypothetical but representative - to illustrate how organizations have dealt with digitization, what the role of leadership has been and what relevant lessons can be learned.

9.1 Case Study 1: Automotive Industry and the Transformation of the Supply Chain

9.1.1 Background and Motivations

A multinational company in the automotive industry, with production plants spread across several continents, identified the need to integrate its manufacturing, logistics and sales processes on a common digital platform. The goal was to reduce operating costs, improve demand forecasting and increase end-customer satisfaction.

- **Problems detected**:
    - Lack of real-time visibility of the supply chain.
    - Manual processes in inventory management, generating errors and delays.
    - Poor communication between manufacturing, logistics and sales areas.

9.1.2 Transformation Strategy

The company designed a three-phase roadmap, aligned with the corporate vision of modernizing the value chain:

1. **Phase I: Digitization of Internal Processes**

- Implementation of an advanced ERP system that centralized production, inventory and sales information.
- Intensive training of middle management and operators to handle the new interfaces and procedures.

2. **Phase II: Automation and Predictive Analytics**
   - Adoption of **IoT technologies** in the manufacturing plant, installing sensors on assembly lines to monitor performance and detect faults.
   - Use of **AI** to forecast demand and optimize inventory, reducing storage costs and lead times.
3. **Phase III: Integration with Customer Experience**
   - Development of a **digital platform** where customers could customize the vehicle, know the production status in real time and request after-sales services.
   - Linking the platform with ERP and predictive analytics systems, allowing information to flow to and from the factory without interruption.

9.1.3 Leadership and Change Management

The digital transformation office (DTO) acted as a catalyst, aligning logistics, production and IT managers. In addition, leadership emphasized the culture of collaboration and the need to share information between departments, providing incentives to those teams that exceeded technology adoption goals.

- **Lewin's model applied**:
  - **Thawing**: Informative sessions on the urgency of being competitive, showing industry cases and market projections.
  - **Change**: Gradual implementation in each plant, with pilot teams disseminating their successes and results to the rest.

- **Refreezing**: Adjustments to operating manuals, performance evaluation linked to digital adoption and an internal recognition system for "innovation champions".

9.1.4 Results and Lessons Learned

- **15% reduction** in logistics costs by synchronizing production with actual demand.
- **20% reduction** in delivery times, thanks to predictive analysis that avoided bottlenecks.
- **Improved customer satisfaction**, valuing transparency in manufacturing and the option to customize their vehicle.

Key lessons:

- **Systems integration** requires a strong management commitment and cultural adaptation of all areas involved.
- **Real-time visibility** (IoT + analytics) multiplies operational efficiency, but requires leadership that promotes cross-functional training and collaboration.
- Including the **customer** in the digital process adds value and builds loyalty, cementing differentiation from competitors.

9.2 Case Study 2: Banking and Process Automation

9.2.1 Context and Objectives

A medium-sized bank with national operations recognized the urgency to modernize its customer service systems and automate back-office processes. Competition from fintechs and growing user expectations for digital services drove the adoption of robotic process automation (RPA) and artificial intelligence technologies for customer service.

- **Challenges**:

- High operating costs in manual procedures (opening of accounts, review of documentation).
- Customers dissatisfied with the slow processing of loans and inquiries.
- Talent with low digital skills, fearful of being left behind.

9.2.2 Digitization Process

1. **Internal Diagnostics**
    - Audit of key processes, identifying opportunities for automation (data verification, consolidation of financial information).
    - Staff surveys on digital skills and readiness to change.
2. **RPA implementation**
    - Bots that handled repetitive tasks in legacy systems, such as form review and data integration across different platforms.
    - 30% reduction in response time for certain back-office operations.
3. **Customer Service through Conversational AI**
    - Chatbots and virtual voice to resolve frequent queries, freeing up human agents for complex cases or cross-selling.
    - Integration of messaging platforms to reach younger audiences.

9.2.3 Talent Management and Training

Resistance to change was mitigated with an **internal training program** in short modules:

- **Automation Fundamentals**: Showing that bots ease the routine burden, making room for higher-value tasks.
- **Data literacy**: Analysis of bot activity reports and proposals for improvement.

- **Upskilling**: Some administrative positions became "automation specialists," trained to configure and monitor bots.

9.2.4 Main Achievements and Lessons Learned

- **25% increase** in customer satisfaction, evaluated through post-service surveys.
- **Focusing human talent** on more complex financial advisory services, instead of repetitive tasks.
- **Cultural changes** that revalued continuous training and the perception of technology as an ally.

Key lessons:

- **Process automation** frees resources for tasks of greater complexity and business impact.
- An effective **upskilling plan** reduces the fear of obsolescence and connects the strategic vision with the professional development of employees.
- **Chatbots** and virtual channels improve customer satisfaction, but require constant monitoring of interaction quality and corrections for possible biases in AI.

9.3 Case Study 3: The Cultural Reconversion of a Publishing Company

9.3.1 Background

A publishing house with several decades in the market, recognized for its print catalog and its network of physical bookstores, faced the disruption of online book sales and the rise of e-books. Its executives decided to **redefine the business model**, exploring digital platforms and reconfiguring the value chain.

- **Central problem**:
    - Drop in print sales, especially in international markets.

- Departmentalized structure, with resistance from editors and senior staff to digitization.

9.3.2 Transformation Strategy

The publisher adopted a **design thinking approach** to reimagine its value proposition:

1. **Reader research**
    - Surveys and focus groups in foreign markets, identifying the growing taste for e-books and online sales platforms.
    - Recognition of segments that value high quality physical books or books with special content.
2. **E-commerce Platform Development**
    - Web portal to acquire titles in physical and digital format, with review and social sharing tools.
    - Integration with logistics systems for fast shipments and real-time tracking.
3. **Alliances with Digital Distribution Platforms**
    - Inclusion of the catalog in global portals.
    - Optimized categorization and metadata to improve positioning in online searches.

9.3.3 Cultural and Structural Reassessment

To break down internal resistance, it was promoted:

- **Training in digital marketing** and data analytics for publishers and sales managers.
- **Cross-functional teams** that brought together editors, designers and IT experts, launching collections based on real-time sales data.

- **Changes in performance indicators**: The company stopped measuring only the print run of physical books, and also valued the adoption of new formats and collaboration in digital platforms.

9.3.4 Results and Main Lessons Learned

- **40% increase** in online sales in two years, offsetting the decline in physical bookstores.
- **Greater** publishing **dynamism**: digital launches with lower costs and faster turnaround times.
- **Revaluation of the printed experience**: special collections linked to interactive applications that increased profit margins and loyalty.

Key lessons:

- **Culture** is crucial for digital transformation to revitalize traditional industries.
- **Design thinking** helps to focus innovation on real customer needs.
- Changing **performance indicators** and reinforcing internal training helps employees perceive digitalization as an opportunity.

9.4 Case Study 4: Lessons from Ford and General Electric

9.4.1 Background

Ford and General Electric have stood out as global corporations that have implemented successful digital transformations in their respective industries. These examples are instructive, though not always replicable in smaller companies or in entirely different sectors. Still, their stories provide valuable lessons about the importance of organizational adaptation and the alignment between technology and corporate strategy.

- **Ford**: Recognized for integrating automation solutions and connected technologies into its manufacturing processes, as well as reconfiguring the customer experience through digital platforms.

- **General Electric**: Noted early adopter of industrial IoT and predictive analytics (with initiatives such as GE Digital), driving efficiencies in manufacturing, services and maintenance.

9.4.2 Critical Success Factors

The Ford and General Electric cases show several elements in common:

1. **Transformational Vision from Senior Management**
    - Leadership clearly defined the digital direction and committed resources for training, infrastructure and process redesign.
2. **Alignment with Organizational Objectives**
    - Technology was integrated in a way that reinforced efficiency and competitiveness, avoiding the superficial adoption of solutions just to "follow the trend".
    - Investments were focused on projects with tangible returns and those that strengthen the culture of innovation.
3. **Focus on Culture and People**
    - Recognizing the resistance to change, upskilling and reskilling programs were promoted.
    - Spaces for co-creation and collaboration were created to foster a sense of belonging and "ownership" of the change among employees.

9.4.3 Criticisms and Limitations

Despite its relevance, it should not be assumed that imitating the Ford or GE model guarantees success. As some analysts point out:

- **Size and Resources**: What works in a global corporation with ample investment capacity may not scale in the same way in a small or medium-sized company.

- **Cultural and Market Differences**: Each sector and region presents unique dynamics that may require different approaches to digital transformation.
- **Constant Evolution**: Both Ford and GE have gone through processes of continuous evolution, sometimes with significant setbacks and adjustments along the way.

Nevertheless, these cases underscore the relevance of visionary leadership, well-contextualized strategies and a balanced appreciation of technology and the human factor.

9.5 Chapter reflections

The case studies presented here - from automotive manufacturing to banking to publishing - along with the iconic examples of Ford and General Electric, show the diversity of approaches that coexist under the umbrella of digital transformation. Although each organization grapples with particular challenges, there are common denominators:

- **Leadership** is a determining factor: it communicates the vision, manages resistance and promotes the culture of innovation.
- **Gradual** technology adoption and constant measurement of results ensure consistency and credibility of change.
- **The training and redesign of indicators** generate commitment and a sense of ownership among personnel.
- **Technology must be aligned with concrete** business **objectives**, avoiding superficial "technological leaps" that do not add real value.

These stories of success (and sometimes of setbacks) confirm that digital transformation is not an end in itself, but a path of continuous evolution, marked by the human perspective, collaboration and the ability to adapt to new disruptions. The final chapter will take up the main lines of this work, proposing a synthesis of lessons learned and recommendations for leaders seeking to guide their organizations towards a sustainable and people-centered digital future.

Chapter 10. Future Challenges and Emerging Trends

Digital transformation is a constantly evolving process, driven by rapid technological progress and socioeconomic changes that reshape business models. While previous chapters have described the key elements of leadership, adoption strategies and success stories, it is essential to look at the challenges ahead and the emerging trends that will set the agenda for the next decade. This chapter explores those forces that will drive digital leadership to new horizons, providing insights into how leaders can prepare for a business future characterized by uncertainty and continuous innovation.

10.1 The Technology Horizon

10.1.1 AI and Collaborative Robotics Expansion

Artificial intelligence (AI) has evolved from the automation of specific tasks to systems capable of processing natural language, learning from complex environments and making decisions with a high degree of autonomy. These capabilities are expected to be further integrated into the daily lives of organizations:

- **Explainable AI**: Demands for transparency in algorithms will emerge, requiring machines to "explain" their reasoning to build trust and mitigate potential bias.
- **Collaborative robotics (cobots)**: Human-machine interaction will become more fluid in industries such as advanced manufacturing, logistics and services. The challenge will be to design safe and harmonious workspaces where people and robots join forces.

10.1.2 Metaverse and Extended Reality

The boundary between the physical and the virtual will become blurred with the massive incorporation of virtual reality (VR) and augmented reality (AR) technologies in areas such as training, customer service and remote collaboration . The metaverse, understood as a shared virtual space, offers unprecedented possibilities for interaction:

- **New business models** based on immersive experiences, virtual spaces for meetings and presentations, and even the sale of digital assets (NFTs).
- **Governance** and cybersecurity **challenges**, as the convergence of the physical and digital can expose users to privacy or fraud risks if clear rules are not established.

10.1.3 Quantum Computing

Although still in an exploratory phase, quantum computing promises to solve exponentially more complex computational problems than those approachable with classical computing. This could:

- **Revolutionizing data analytics**: enabling the processing of colossal volumes of information in areas such as pharmacology, finance and logistics.
- **Generate new vulnerabilities** in cryptography, jeopardizing information security if algorithms resistant to quantum computing are not developed.

10.2 The Evolution of Business Models

10.2.1 Digital Platforms and Collaborative Ecosystems

The idea of platforms as a business hub (e.g. marketplaces, social networks, service hubs) will not only continue to grow but will become more sophisticated:

- **Multi-sector ecosystems**: Companies from different sectors will collaborate through shared platforms to offer comprehensive customer solutions.
- **Expansion of "servitization"**: Where once a product was sold, now a continuous service is sold (e.g. software as a service, industrial machinery with intelligent maintenance).

10.2.2 Circular Economy and Sustainability

Social and regulatory pressure to adopt sustainable practices will intensify. Digitalization can act as a catalyst for the circular economy:

- Real-time **life cycle analysis** to minimize the environmental footprint at each stage of production and distribution.
- **New business opportunities** linked to the reuse, recycling and sharing of assets, enabled by technological platforms.

10.2.3 FinTech and Financial Decentralization

Financial decentralization (DeFi) and the rise of technologies such as blockchain will continue to reshape banking and economic transactions:

- **Global payments and smart contracts** to eliminate intermediaries, accelerate liquidity and reduce costs.
- **Digital microfinance** that empowers sectors underserved by traditional banking.
- **Dynamic regulation** to balance innovation with user protection, anti-fraud controls and monetary stability.

10.3 Organizational and Leadership Challenges

10.3.1 Comprehensive Cybersecurity Management

The more processes are automated and interconnectivity grows, the greater the exposure to cyber-attacks. Leaders must evolve toward a holistic security approach that considers not only the technical aspects, but also:

- **Ongoing** employee **training** in secure practices (passwords, phishing, ransomware).
- **Contingency** and cyber resilience **plans**, contemplating total disruption scenarios and rapid recovery strategies.

- **Ethics and responsibility** in the handling of confidential data, respecting the privacy of clients and collaborators.

10.3.2 Digital Talent and Skills Gaps

The demand for digital talent (data scientists, AI specialists, cloud solution developers, etc.) will outstrip the supply available in many labor markets, generating salary pressures and fierce competition to recruit the best profiles.

- **Internal retention and development strategies**: Organizations that build and scale their own teams will have an advantage in reducing dependence on outsourcing.
- **Diversity and inclusion**: Seeking talent from underrepresented populations can be a way to alleviate the shortage of specialists while promoting an innovative culture.

10.3.3 New Psychological Contract with Employees

The fourth industrial revolution brings with it uncertainty about job stability, the robotization of tasks and the adoption of agile methodologies that disrupt the concepts of hierarchy and career. Leadership will have to:

- **Reconfigure the value proposition** to employees, offering flexible career plans, continuous training and more collaborative and autonomous work environments.
- **Promote well-being and cohesion** in more dispersed, hybrid or remote organizations, avoiding disconnection or digital burnout.

10.4 Towards an Ethical and Common Good Oriented Leadership

10.4.1 Social Responsibility and Algorithmic Governance

Massive digitization and AI raise unprecedented ethical implications. In the near future, we will see the demand for leadership capable of:

- **Regulate** the design and use of algorithms with a view to fairness, non-discrimination and transparency.

- **Balance** the interests of the company with the protection of personal data and the dignity of users or customers.
- **Promote sustainability** and the common good, ensuring that the benefits of technology are distributed equitably and do not aggravate social gaps.

10.4.2 Reconfiguring the Human Dimension

Paradoxically, as technology advances, the human factor becomes more valuable in areas such as creativity, complex decision making, empathy and interdisciplinary collaboration. The leader of the future must:

- **Cultivate emotional intelligence** in the organization, creating environments of trust and autonomy that foster innovation.
- **Inspire with vision**: When automation is capable of executing a large part of the tasks, the leader's role is to provide direction and a sense of purpose that motivates people beyond the purely instrumental.

10.5 Chapter Reflections

The outlook for the coming decades combines dramatic technological breakthroughs with daunting ethical and organizational challenges. AI, extended reality, quantum computing, and new business models based on platforms and decentralization promise to radically transform industries. However, such transformations require leadership capable of balancing efficiency with equity, innovation with responsibility, and immediacy with a long-term perspective.

In this sense, the adoption of emerging technologies cannot be viewed in isolation, but rather as part of a comprehensive strategy where organizational culture and ethics are central. Leaders who understand this reality and are willing to adjust power dynamics, work structures, talent development and business models will be more likely to successfully navigate the digital future.

In light of the previous chapters, the importance of global leadership that combines digital competencies, emotional intelligence, critical thinking and an inclusive approach is reaffirmed, recognizing that constant change is the new normal. With accountability and collaboration as guides, organizations can turn these challenges into opportunities, driving sustainable growth and a digital transformation that is humanized and consistent with societal values.

The path to digital maturity involves continuous exploration, a willingness to experiment and learn, and the firm conviction that people, when empowered and valued, are the most strategic asset in the age of automation and artificial intelligence.

Chapter 11. General Conclusions and Recommendations

The digital era has brought with it profound changes in business models, organizational culture and the way leadership is exercised. Throughout the chapters, key elements for understanding and successfully guiding **digital transformation** have been addressed: from the importance of strategic vision and change management, to the need to balance the adoption of technologies with the preservation and growth of human capabilities. This **final chapter** provides a synthesis of the main findings and puts forward concrete **recommendations** for leaders and organizations seeking to thrive in a highly competitive and technologically-driven global environment.

11. Main Conclusions

11.1 Digital Transformation is an Integral Process

Process digitization, AI implementation and automation cannot be seen as isolated projects, but as part of a comprehensive process that encompasses strategy, culture and the experience of employees and customers. This global vision is key to aligning technology efforts with core business objectives.

11.1.2 Global Leadership Demands New Competencies

Leaders in the digital age must develop hybrid leadership, combining:

- Digital skills and understanding of emerging technologies.
- Emotional and social skills (empathy, communication, motivation).
- Adaptability, creativity and strategic thinking.

Digital transformation is no longer a purely technological issue but an integral leadership challenge, where vision and change management are as important as the adoption of new tools.

11.1.3 Organizational Culture is the Substratum of Innovation

Culture is the factor that can accelerate or hinder the adoption of innovations. An organization that values collaboration, continuous learning and trust among its members will be better prepared to assimilate the changes brought about by digitalization. However, establishing or reinforcing this culture requires:

- Committed and exemplary leadership.
- Training and support programs.
- Structural changes that reflect and sustain the new practices.

11.1.4 Resistance to Change is Natural, but Not Insurmountable

Resistance surfaces when people fear losing their role, their relevance or the security of the routines they dominate. Lewin's Organizational Change Model, along with approaches such as Bass' Transformational Leadership, show that unfreezing the mindset, introducing change with accompaniment and consolidating new practices (refreezing) are essential steps. Effective communication, active participation and adequate training significantly decrease this resistance.

11.1.5 Ethics and Sustainability Emerge as Imperatives

The massive use of data, automation and the convergence of technologies (AI, IoT, metaverse, etc.) pose ethical and social dilemmas that leaders must address responsibly. Algorithmic responsibility, privacy protection, digital inclusion and environmental impact are aspects that can no longer be ignored if we aspire to a humanized and sustainable digital transformation.

11.2 Recommendations for Practice

11.2.1 Develop a Strategic and Flexible Roadmap

- **Clarify** transformation **objectives** (efficiency, innovation, diversification, etc.).
- **Stagger the adoption of technologies** in phases, obtaining "quick wins" that generate internal credibility and motivation.
- **Periodically review** the plan to adjust priorities according to market changes or technological advances.

11.2.2 Strengthening Hybrid Leadership

- **Invest in continuous training** for managers and middle management, combining digital competencies and soft skills (emotional intelligence, creativity, communication).
- **Promote cross-mentoring**, where technology experts collaborate with traditional leaders to exchange knowledge and approaches.
- **Encourage self-leadership**: empower teams with autonomy and empower the "champions" of digital transformation.

11.2.3 Involving Employees in the Process of Change

- **Co-create solutions** with end users: design thinking sessions, cross-functional innovation teams and pilot projects that gather feedback from staff.
- **Communicate the purpose** behind each initiative, linking digitalization to employee well-being, company growth and customer satisfaction.
- **Design reskilling and upskilling plans** that strengthen employability and reduce anxiety in the face of automation.

11.2.4 Establishing Clear Technology Governance

- **Form innovation committees** or create a Digital Transformation Office (DTO) to oversee the consistency and quality of initiatives.
- **Monitor cybersecurity** and data protection through updated policies and protocols, raising awareness throughout the organization on secure practices.
- **Define success indicators** in the economic, process and cultural areas (customer satisfaction, organizational climate, ROI of digital projects).

11.2.5 Cultivating Ethics and Sustainability

- **Adopt an algorithmic governance approach**, reviewing the transparency and fairness of AI models.

- **Evaluate the environmental impact** of the technologies implemented, considering, for example, the energy efficiency of the data centers.
- **Encourage the inclusion of** people with diverse backgrounds and skills, recognizing that the plurality of perspectives enhances innovative capacity and social responsibility.

11.3 A Vision for the Future of Global Leadership

The final reflections point to the continuity of the digital transformation as an unfinished phenomenon, which will evolve with the emergence of new technologies (quantum computing, metaverse, advanced robotics) and with the social and economic changes that will redefine human needs. In this context, global leadership will take on even greater challenges for:

1. **Balance innovation with human dignity**: Digital tools should empower human talent, not displace or dehumanize it.
2. **Maintain cultural coherence**: Ensure that technology is used for purposes aligned with the organizational mission, taking care of internal cohesion and the relationship with the community.
3. **Deepen inter-organizational collaboration**: The complexity of the digital environment invites companies to form alliances, share knowledge and co-create solutions that transcend sector and geographic boundaries.

By following the recommendations and lessons learned throughout this book, leaders will be able to lay the foundations for a sustainable digital transformation, where technological innovation and human development are mutually reinforcing. Far from being a point of arrival, digitalization stands as a path of continuous adaptation, where each step requires a combination of intelligence, ethics and shared vision.

To the extent that organizations manage to humanize automation, co-create with their collaborators and orient technology to the common good, they will be prepared to face the challenges ahead with determination and resilience.

Because, ultimately, success in the digital age lies in the ability to lead with purpose, inspire team trust and commit to progress that is genuinely inclusive and enriching for people and for society as a whole.

References

Asatiani, A., Copeland, O. & Penttinen, E. (2023). Deciding on the robotic process automation operating model: A checklist for RPA managers. *Business Horizons, 66*(1), 109-121. .

Avolio, B. J. & Kahai, S. S. (2003). Adding the "E" to E-leadership:: How it may impact your leadership. *Organizational dynamics, 31*(4), 325-338. https://doi.org/10.1016/S0090-2616(02).

Bass, B. M. & Riggio, R. E. (2006). *Transformational leadership* (2nd ed.). Psychology Press.

Chamorro-Premuzic, T. (2021). The essential components of digital transformation. *Harvard Business Review, 13*, 1-6. https://hbr.org/2021/11/the-essential-components-of-digital-transformation.

Chui, M., Manyika, J. & Miremadi, M. (2016). Where machines could replace humans-and where they can't (yet). *The McKinsey Quarterly*, 1-12.

Chui, M., Manyika, J., Miremadi, M., Henke, N., Chung, R., Nel, P. & Malhotra, S. (2018). *Notes from the AI frontier: Insights from hundreds of use cases*. McKinsey & Company. https://www.

Cortellazzo, L., Bruni, E. & Zampieri, R. (2019). The role of leadership in a digitized world: A review. *Frontiers in psychology, 10*, 456340. https://doi.

Deloitte Insights (2020). *2020 global technology leadership study*. Deloitte Touche Tohmatsu Limited. https://www2.

Deng, C., Gulseren, D., Isola, C., Grocutt, K. & Turner, N. (2023). Transformational leadership effectiveness: an evidence-based primer. *Human Resource Development International, 26*(5), 627-641. https://doi.org/10.1080/13678868.2022.2135938

Endrejat, P. C. & Burnes, B. (2024). Draw it, check it, change it: reviving Lewin's Topology to facilitate organizational change theory and practice. *The Journal of Applied Behavioral Science, 60*(1), 87-112.

World Economic Forum (2021a). *Global technology governance report 2021*. .

World Economic Forum (2021b). *Harnessing technology for global goals.* https://www3.

Hoffmann, V. (2007). Book Review: Five editions (1962-2003) of Everett ROGERS: Diffusion of Innovations. *The Journal of Agricultural Education and Extension, 13*(2), 147-158.

Kane, G. C., Palmer, D., Phillips, A. N., Kiron, D. & Buckley, N. (2015). *Strategy, not technology, drives digital transformation.* MIT Sloan Management Review. https://sloanreview.

Khan, S. (2016). Leadership in the digital age: A study on the effects of digitalisation on top management leadership. Stockholm Business School

Klus, M. F. & Müller, J. (2021). The digital leader: what one needs to master today's organizational challenges. *Journal of Business Economics, 91*(8), 1189-1223.

Mendenhall, M. E., Reiche, B. S., Bird, A. & Osland, J. S. (2012). Defining the "Global" in Global Leadership. *Journal of World Business, 47*(4), 493-503.

Powell, B. J., Beidas, R. S., Lewis, C. C., Aarons, G. A., McMillen, J. C., Proctor, E. K. & Mandell, D. S. (2017). Methods to improve the selection and tailoring of implementation strategies. *The journal of behavioral health services & research, 44*(2), 177-194.

Rogers, E. M. (2003). *Diffusion of innovations* (5th Ed). Free Press.

Rogers, E. M., Medina, U. E., Rivera, M. A. & Wiley, C. J. (2005). Complex adaptive systems and the diffusion of innovations. *The innovation journal: the public sector innovation journal, 10*(3), 1-26.

Ross, J. W. & Beath, C. M. (2002). *Beyond the business case: New approaches to IT investment.* MIT Sloan School of Management, Center for Information Systems Research.

Ross, M. & Taylor, J. (2021). Managing AI decision-making tools. *Harvard Business Review, 1*(1), 11-27. https://hbr.org/2021/11/managing-aidecision-making-tools.

Sambamurthy, V., Bharadwaj, A. & Grover, V. (2003). Shaping agility through digital options: Reconceptualizing the role of information technology in contemporary firms. *MIS quarterly, 27*(2), 237-263. .

Schwartz, T. & McCarthy, C. (2007). Manage your energy, not your time. *Harvard business review, 85*(10), 63.

Sheninger, E. (2019). Digital leadership: Changing paradigms for changing times. Corwin Press.

Van Oorschot, J. A., Hofman, E. & Halman, J. I. (2018). A bibliometric review of the innovation adoption literature. *Technological Forecasting and Social Change, 134*, 1-21.

Vanderslice, S. (2000). Listening to Everett Rogers: Diffusion of innovations and WAC. *Language and Learning Across the Disciplines, 4*(1), 22-29.

MIX
Papier aus verantwortungsvollen Quellen
Paper from responsible sources
FSC® C105338

Printed by Books on Demand GmbH, Norderstedt / Germany